Two Studies in Logic

Valentin B. Bura

THE ARGUMENT FOR DIALETHEISM FROM GÖDEL'S INCOMPLETENESS THEOREM

VALENTIN BURA

ABSTRACT. Dialetheism is a relatively recent intellectual product, which owes most of its present shape to the work of Graham Priest. He is known as a strong believer in the existence of true contradictions. In this paper, we are concerned with the Argument for Dialetheism as given by Professor Priest in the third chapter of [1]. We closely outline and discuss the argument together with the possible objections identified in the book. We analyse the author's commitments and the main points that make the deployment of Godel's result possible. Our main contribution is showing that the argument is fallacious, by analyzing the use of Gödel's First Incompleteness Theorem, found in the book. The argument in the book fails to give a correct Interpretation to the conditions required for the Theorem to be legitimately applied. We show the correct line of reasoning and argue that Godel's result cannot be used in this context. We continue with criticizing the author's view that the informal Mathematical English used in proving Theorems of Mathematics could be formalized. We claim that the author is committed to a view of Mathematical Practice as automatizable, at least in principle. We use an Information-Theoretic argument as found in [5] to show that this view cannot be sustained.

1. INTRODUCTION

Paraconsistent Mathematics comes in two flavours: as advocated by Graham Priest in many of his writings, of which representative is "In Contradiction", and as outlined by Chris Mortensen in [13]. The motivation of Priest is slightly theoretical, the logical paradoxes, while Mortensen's 'hands-on' approach tries to effectively model inconsistent data-sets.

The first three chapters of [1] are set to give philosophical motivations for the existence of dialetheias, statements that are both true and false. The first two chapters deal with semantic and set-theoretic paradoxes, while the third focuses on devising an argument using Gödel's First Incompleteness Theorem.

There are several constructions made to avoid paradoxes which orthodox Logic and Mathematics have come to accept, like Tarski's Semantic Hierarchy, described in [11], and the Set Theoretic Cumulative Hierarchy introduced by Von Neumann. Professor Priest criticizes them vehemently, also dismissing attempts for consistency as in [12], while he comes to establish a trade-off between consistency and completeness in a formalized theory.

However, the arguments employed in the first two chapters are more informal and appeal towards persuading the reader, rather than proving beyond any doubt that Dialetheism must be true. They amount to saying that certain structures do not faithfully represent our 'naive', unguided intuitions, and so their raison d'être is purely the need for consistency, which cannot, by itself, suffice. This orthodox need for consistency would presumably be nothing but a prejudice and could reasonably

Date: January 7, 2012.

be sacrificed if we are to be faithful in representing our semantic/mathematical concepts.

The third chapter is the one which, in our opinion, gives the strongest argument for the existence of dialetheias. It is not argued that we ought to sacrifice consistency, like in the previous two chapters; rather it is shown that, at least in our informal mathematical discourse, dialetheias must exist.

Hence if we want to stay committed to a consistent theory of Mathematics, we must thoroughly analyse the argument set out in the third chapter of "In Contradiction".

2. The Theorem

Priest uses Gödel's First Incompleteness Theorem to try to show that there must exist dialetheias, statements which are both true and false in a formal system. This argument is part of a larger one that deals with motivations for Dialetheism, which establishes a very important trade-off between consistency and completeness - namely, the conclusion is that if we wish to maintain consistency, we must do it with the cost of sacrificing completeness, or rather that a consistent system will necessarily be incomplete.

The statement of the Theorem, as given by Professor Priest, is as follows.

Theorem 1 (Gödel's First Incompleteness Theorem). *Let T be a theory which can represent all recursive functions and whose proof relation is recursive. Then there is a formula ϕ such that:*

(i) If T is consistent, ϕ is not provable in T;

(ii) If the axioms and rules of T are intuitively correct, we can establish by an intuitively correct argument that ϕ is true.

In this context, "recursive" is equivalent to "effectively recognizable" or "computable". A structure is recursive, whenever we have an effective procedure/algorithm for producing it. To say that the Theory can represent all recursive functions is to say that it is strong enough to have the ability to do basic arithmetic and represent arithmetical truths.

The Gödel sentence ϕ will then be of the form $\neg \exists x \Pi(x, n')$. Here $\Pi(x, y)$ is a formula with two free variables whose sense is that x is the code of a proof of formula with code y, so in the original sentence, n is the code of ϕ and n' is its numeral. In effect, the original formula is equivalent to the statement "I am not provable", reminiscent of the celebrated Liar Paradox.

The proof of the theorem, also given in [1] is as follows.

Proof. Given a theory T that satisfies the conditions of the theorem, we code each formula and proof of T as a number. This is standardly done using the well-known technique of Gödel numbering. We skip the details here.

If α is a formula, let $\#\alpha$ be its code and let $\#\alpha'$ be the numeral of its code.

If $\alpha(v)$ is a formula of one variable, its diagonalization is $\alpha(\#\alpha(v))$. We note that, since diagonalization is a recursive procedure, it is representable in T.

Then there is a term of the language, $\delta(x)$, of one free variable, such that if m is the code of the diagonalization of the formula with code n, $T \vdash \delta(n') = m'$. In other words, δ "computes" the diagonalization code of its argument.

Lemma 2. *If $\alpha(v)$ is any formula with one free variable v, then there is a sentence β such that $T \vdash \beta \leftrightarrow \alpha(\#\beta')$.*

Proof. We consider the formula $\alpha(\delta(v))$ of code n, hence of diagonalization $\alpha(\delta(n'))$. Let the diagonalization have code k. We know that $T \vdash \delta(n') = k'$. Hence $T \vdash \alpha(\delta(n'))) = \alpha(k')$. Thus $\alpha(\delta(n'))$ is the required sentence.

$\square$

Now, given a pair of numbers (m, n) we could effectively tell whether m is the code of a proof of a formula with code n. By representability, we can find a term of the language that characterizes this relationship. Let $\Pi(x, y)$ be this term.

Hence, if m is the code of a proof of a formula with code n, $T \vdash \Pi(m', n') = 1$ otherwise $T \vdash \Pi(m', n') \neq 1$.

Consider the formula $\neg \exists x \Pi(x, y)$. By the Lemma, we can find a formula ϕ such that $T \vdash \phi \leftrightarrow \neg \exists x \Pi(x, \#\phi')$.

Suppose that $T \vdash \phi$. Then $T \vdash \neg \exists x \Pi(x, \#\phi')$. But some m is the code of the proof of ϕ. Hence $T \vdash \exists x \Pi(m', \#\phi')$ and therefore $T \vdash \exists x \Pi(x, \#\phi')$. So T is inconsistent. Therefore, if T is consistent, then it is not the case that $T \vdash \phi$. So ϕ is not provable in T.

The second part of the theorem states that if T is sound then ϕ is true. We now prove this.

If ϕ is provable then by soundness it is true. If ϕ is not provable, then no number is a code of its proof. Hence $\forall x \neg \Pi(x, \#\phi')$ which is equivalent to $\neg \exists x \Pi(x, \#\phi')$ which is equivalent to ϕ, hence ϕ must be true.

To complete the proof, we need an argument that shows that T is sound if its rules are intuitively correct. We skip this step, as the proof is trivial and mechanical.

$\square$

3. The Argument for Dialetheism

The argument then goes on like this.

We have established using a fragment of informal Mathematical English that ϕ is both true and unprovable in the formal system.

Now, mathematicians seem to agree that their informal language can be, in principle and if required, completely formalized...

Let T be the formalization of our naive proof procedures. It is contended that T satisfies the conditions of Gödel's Theorem. Then, by this theorem, if T is consistent, there is a statement ϕ which is not provable in T which we can establish by an informal argument that it is true, hence it is provable in T, since T is precisely the formalization of our naive procedure. Hence ϕ is both provable and not provable in T.

We obtain a contradiction, and hence by a reductio argument we must conclude that our naive proof procedures are inconsistent.

There is an alternative version of the argument that avoids the reductio: if ϕ is provable, then T is inconsistent; ϕ is provable; therefore T must be inconsistent.

But these proof procedures are the means by which we establish theorems as being true or false. It follows that some contradictions are true, hence Dialetheism is correct.

4. Possible Objections as Anticipated by Priest

The possible objections to this use of Gödel's Incompleteness Theorem as discussed in "In Contradiction" could be:

(i) Mathematical English cannot be formalized;

(ii) The formalized theory could not represent all recursive functions;

(iii) The naive proof relation is not recursive, as possibly argued by Intuitionists, Anti-mechanists and people who argue that the formalization must necessary be diachronic, rather than synchronic.

It seems to be beyond doubt in Priest' account that the fragment of Mathematical English used on proving Gödel's Theorem could be, in principle, formalized. This means that its syntax would be carefully re-arranged so to make it a formal language and its set of theorems would be deductively closed. Indeed, we have to subscribe to this position if we are to believe that Mathematics is anywhere close to the province of Logic. And by all existing evidence and philosophical intuition, there is no other way of doing Mathematics than by obeying basic mechanisms of Logic!

There is then no doubt that this formalized theory could represent all recursive functions, since it is possibly stronger than the original formal system for doing arithmetic, which could, in effect, represent all recursive functions. This as much seems to be beyond doubt.

However it can be doubted that the proof relation of the theory is recursive. Priest responds to this by arguing that it is "part of the very notion of proof that a proof should be recognizable as such". And if the proof relation is effectively recognizable then by Church's Thesis it must be recursive.

First of all, it is noted that Intuitionism doesn't provide an objection to the claim that proof is recursive since this account is that "truth is not effectively recognizable while proof is".

Secondly, there is the so-called Anti-mechanist account which also tries to establish the non-recursiveness of the notion of truth, in which Gödel's Theorem is invoked as follows:

"Let T be a Turing machine which 'represents' a human mathematician in the sense that T can prove just the mathematical statements they can prove. Then using Gödel's technique, a human mathematician could discover a proposition that T cannot prove, and moreover she can prove this proposition. This refutes the assumption that T 'represents' our naive proof mechanism; hence we are not a Turing machine. Hence, the mechanism of proof that we are using is profoundly different from a Turing Machine, which means it is not recursive."

In this setting, the Church-Turing thesis is invoked in the assumption that a formal system is equivalent to a Turing Machine. This line of reasoning is sometimes known as the "Mathematical Objection" or the "Gödel Argument", and has been used, amongst others, by Lukas and Penrose to argue that Mathematical thinking is not algorithmic and even that the human mind could not be simulated by a Turing Machine. This intimate relationship between Gödel's Theorem and human creativity is still a much debated topic and it remains an open question.

Priest' response is to underline that the argument requires the assumption that the human mind/formal system/Turing Machine is a consistent system deductively - for only then can the Incompleteness Theorem be applied to it. And this, Priest argues, is question begging in this context, since inconsistency is precisely the stake in play. We will return to this line of reasoning later, however we note that the dialetheist position can be defended here reasonably well: it is more than reasonable to suppose that the human mind might be, as a matter of fact, inconsistent since it is prone to making mistakes. Indeed, we make mistakes all the time and our inferences will always be subject to revision - this is perfectly consistent with the assumption that our mind is a formal mechanism and that our proof mechanisms are recursive. This is related to the claim of 'strong' Artificial Intelligence, namely, that the human brain can be in principle simulated by a Turing Machine.

The third and most important objection to the recursiveness of naive proof is the suggestion that, from time to time, our axioms and rules of inference might change in a non-rule-governed fashion. Hence, we may arbitrarily choose which rules of inference and axioms to add based on empirical considerations - Gödel suggested that new axioms might be accepted based on the fact that they imply many things already known to be true and nothing known to be false.

This objection could only stand if it can be shown that there is a need to step into a process of revision of our axioms or rules of inference for the original argument to work; otherwise the dialetheist can be quite content that he has theoretically devised a static system, which models a fragment of Mathematical English and which is inconsistent.

Therefore, for the sake of fairness, it must be said that the onus of proof is on those doubting the dialetheist argument to try to show that such a continous revision must be pursued.

Our system would then 'grow' in two directions - forward, towards new theorems, and backwards, towards the foundations, by specifying/revising axioms and possibly the rules of deduction. This is to say that a formal coding of the naive proof mechanism would necessarily be diachronic, not synchronic, and the formulation and proof of the Gödel sentence would launch us in a process of revision of the proof procedures in force.

One suggestion is that the proof relation is changed by the addition of the Gödel sentence as a new axiom. The dialetheist response here is that the Gödel sentence was, as a matter of fact, provable in the first place and that whatever changes to the system, they would not be arbitrary but would follow directly from the old setup of the system. An idea is that new vocabulary and concepts might be added indefinitely, like the property of being a true statement of the system, as suggested by Dummett. This is easily refuted by Priest, who argues that this property is recursive as such.

Priest is ready to grant the supposition that the proof relation might be a diachronic one, however in this case he maintains that the transition to a new system from the old one is rule-governed as well - in effect we not only have rules of proof for generating theorems, but rules for generating rules of proof for theorems, in a completely deterministic, rule-governed fashion.

Priest then invokes Craig's Theorem, which states that any recursively enumerable set of well-formed formulas of a first-order language is (primitively) recursively axiomatizable. This is not accurate as such, since the first-order condition of the theorem is not satisfied in this context.

5. A Discussion of the Argument

Gödel's Incompleteness Theorems are widely discussed in Philosophy, in contexts that range from Epistemology to Philosophy of Mind. Philosophers love using them in different arguments, while mathematicians have grown suspicious to this usage. Everybody agrees, however on the profoundity of Gödel's result.

It is somehow intuitive for the theorem to be used in this context, since its corollary, the Second Incompleteness Theorem would seem to guide us into this line of reasoning.

We could, as a matter of fact, trace the roots of Priest' argument to this corollary of Gödel's main theorem. Let us have a look at this result.

Theorem 3 (Gödel's Second Incompleteness Theorem). *In any consistent axiomatizable theory which can encode sequences of numbers the consistency of the system is not provable in the system.*

Proof. Let T be such a theory. Let ϕ be its Gödel sentence. We will refer to the property of consistency of T as $Con(T)$.

Now, suppose T proves $Con(T)$, then T proves ϕ.

But T also proves that it doesn't prove ϕ. Contradiction.

Hence T cannot prove its own consistency. □

The theory, therefore, cannot prove its own consistency. Could it be, then, that we can legitimately apply this line of reasoning to our informal proof procedures? This is what the dialetheist argument does and it resembles quite well the proof outlined above.

Essential to the dialetheist argument is the possibility of a certain "collapse" of our naive proof procedures into one single formalizable theory. Such a theory would completely specify the practice of mathematical argumentation, as it is used at one given moment in time.

Therefore, one important commitment that the dialetheist makes is to a view of the mathematical practice as possible to be simulated by a Turing Machine. (This is so because of the Church-Turing Thesis that equates the class of recursive and computable functions.) This machine would be, if we follow the argument, either 'static' or 'dynamic', capable of evolving towards a deeper and deeper 'understanding' of Mathematics through specifying new Axioms.

It must be said that the argument does not apply to the current foundation for Mathematics, the Zermelo-Fraenkel Set Theory with Choice (ZFC). This is because our naive proof procedures transcend ZFC in power. Mathematicians like to say that all of their theorems are theorems of ZFC, but this is not actually accurate. Few examples are statements like "ZFC has a countable model" or "the constructible universe is a model of ZFC" and similar. These are 'theorems', but they can't actually be theorems of ZFC, since they imply the consistency of ZFC. They are, however, theorems of ZFC+Con(ZFC).

The formalized theory obtained would then presumably be equivalent to a collapsed Feferman hierarchy as described in [6]. We will encounter Feferman's construction later on in our analysis.

It is then the dialetheist position that the incompleteness phenomenon discovered by Gödel is due to our theories being consistent. In all fairness, professor

Priest confesses agnosticism to the possibility that an inconsistent non-trivial formal system could be complete, however it will be straight-forward for us to show, using Information Theoretic arguments, that this cannot be the case.

Even though Priest is ready to accept the possibility that the formalization of Mathematics will have to be diachronic, we can argue for this as follows.

If we grant that the dialetheist view is correct, that there are true contradictions within the discourse of Mathematics, it could be the case that this would only be pathological and that the various deduction systems employed could be 'patched' to achieve consistency, perhaps at the sake of completeness. Mathematicians would then enter an ongoing debate about which axioms to keep and which to sacrifice, much of what is debated nowadays regarding acceptance of axioms.

If the Dialetheist position would be proven correct, perhaps this would result in the impossibility to specify a unified set of axioms for the whole of Mathematics, and the work of logicians would from then on be concerned with drawing and understanding satisfactory consistent foundations to each mathematical area, much like the modern project of Reverse Mathematics. The axiomatic foundations would then be diachronic in Priest' understanding of the concept.

In what follows, we will endeavour to show that the dialetheist argument is actually mistaken and that Priests' related argumentation that the practice of Mathematics must be recursive cannot be reasonably sustained.

6. The Dialetheist Argument is Mistaken

Our proof that the argument is incorrect is not hard to digest. As a matter of fact, it reduces to observing a discrepancy between what Priest claims he can show and what can actually be shown.

Let T be our formalized theory, let ϕ be its Gödel sentence, let $Con(T)$ be the statement that T is consistent and let $Prov(T, \phi)$ be the statement that T proves ϕ. What can be shown in this context is strictly that $T \vdash Con(T) \to \phi$, and hence only that $T + Con(T) \vdash \phi$, also that $T + Con(T) \vdash \neg Prov(T, \phi)$.

Hence we have, by Gödel's argument, that $T \nvdash \phi$ and that $T + Con(T) \vdash \phi$, which is far from contradictory. This is the place of the important assumption made by the Incompleteness Theorem that the system must be consistent.

In effect, we have the two statements $T + Con(T) \vdash \neg Prov(T, \phi)$ and $T + Con(T) \vdash Prov(T + Con(T), \phi)$. This is the pair that would supposedly produce a contradiction.

The reason why we need to assume that the axioms and rules of T are "intuitively correct" in the intuitive proof that ϕ is true (in Priest' setup of the Theorem) is that we need them at least to be consistent (in fact, ω-consistent, although this assumption can be removed by using Rosser's argument). So when we do formalize the intuitive argument that ϕ is true, what we get is a formal proof that 1) if T is consistent, then T does not prove ϕ; but 2) T does prove "$Con(T) \to \phi$," where $Con(T)$ is some formalization in T of the statement that T is consistent!

Crucial to the obvious conclusion that the argument must be mistaken is the following remark: ϕ only says of itself "I am not provable in T" and makes no commitment whether or not it is provable in $T + Con(T)$. Well it can hardly say anything about $T + Con(T)$ since this is an entirely new formal system, and a new aritmethization procedure must be carried out, in order to generate a new Gödel sentence ϕ' which says "I am not provable in $T + Con(T)$" and so on...

The dialetheist claims that we can collapse this hierarchy and obtain a system which can prove its own Gödel sentence. But the only way in which it could it is by proving its own consistency, which would deem it inconsistent, as shown above. Hence any approach of this kind would be necessarily begging the question.

In all fairness, the dialetheist argument acknowledges the needed assumption of $Con(T)$, however it fails to take into account that $T + Con(T)$ transcends the original system. In effect, the argument claims to produce the contradiction that $T \nvdash \phi$ and $T \vdash \phi$ which is not at all what can actually be shown.

7. The Transfinite Progression

What then, can we make of this attempt of arguing for true contradictions from Gödel's Incompleteness Theorems? Can the approach be re-used in perhaps a different form to argue for the same conclusion? We have seen where the mistake is, what happens if we patch our logic somehow?

We have established that, starting from a base theory T, which can be the Peano Arithmetic or something even stronger, we can show there is a statement ϕ of T which we can prove is true and unprovable from T, provided we have $Con(T)$. (The base theory could be any theory of our choice - it is sensible to start from PA for simplicity, but it can also be the theory of naive mathematical proofs.)

Now what happens if we add $Con(T)$ to T in order to incorporate ϕ into the theorems we can actually derive? In this case we obtain a strictly stronger theory T', which includes T and all of its theorems but can prove much more than T.

Can we successfuly apply the Dialetheist argument to T'? The answer is 'No', because we still do not know whether or not T' is consistent. In effect, we would have a new Gödel sentence ϕ', and we could transcend the system again by adding $Con(T')$ as an axiom.

The construction can be carried on transfinitely as follows.
(i) T_0 is given;
(ii) $T_{k+1} = T_k \cup Con(T_k)$;
(iii) for k a limit ordinal, $T_k = \cup_{i<k} T_i$.

Two questions immediately arise in this context. The first one is what exactly is the proving power of each ordinal and, secondly, provided at some limit ordinal we could prove all of the true statements of arithmetic (for example), could we in effect collapse the hierarchy into one single recursively enumerable set of axioms? This seems to be the necessary condition for the Dialetheist argument to work, given the circumstances - that is, we need a theory that can prove its own Gödel sentence.

First of all, it should be noted that that the construction of theories may depend on the notation we use for an ordinal. In particular, Gödel's construction only works if the theory is recursively enumerable. Only then we can formalise the notion of provability from a theory T in a base theory such as PA. In other words, the provability predicate depends on a recursively enumerable index for T.

At the limit stages of our hierarchy, to get a recursively enumerable index for the theory, we need to provide an effective list of the theories whose union we are taking. This will invariably depend on the computable presentation of the ordinal. In particular, it can only be done for computable ordinals. In effect, the construction cannot go beyond ω_1^{CK}.

What is even more discouraging is that at no stage we will get the full theory of the natural numbers (true arithmetic), since that theory is very far from being recursively enumerable. So the answer to our second question seems to be 'No'. That is, we cannot obtain a computable theory of 'True Arithmetic', a predictable answer given Gödel's Incompleteness Result.

Regarding our first question, Feferman in [6] has used, following the work of Turing, a slightly different hierarchy where instead of $Con(T_k)$ he added a soundeness statement $Prov_{T_k}(\phi) \to \phi$. What he was able to show was that a sequence of length ω^{ω^2+1} can be extracted from our hierarchy, such that every theorem of True Arithmetic is provable in some theory of the sequence. This result is known as the Feferman Completeness Theorem, which is explained in detail in [7].

This result, however, is weak and by no means it implies that every truth is attainable in Mathematics, let alone that a mechanical procedure, as complex as imaginable, could in principle enumerate the Theorems of, say, True Arithmetic. In the words of Torkel:

"It would be misleading to say that [Feferman's] completeness theorem shows that we, or an idealized mathematician, will 'eventually' obtain every arithmetical truth by iterating reflection principles, since completeness depends essentially on a very careful choice of path in the set O [described by Kleene] of ordinal notations. (Indeed in Feferman and Spector [1962] it is shown that there are paths through O - paths giving a notation to every constructive ordinal -such that the corresponding sequences derived from any of a wide category of families of theories do not even prove every true Π_1-sentence.) It may also be a bit misleading to speak of a 'choice of path' here, suggesting as it does that by intuitively sniffing out the right choice at each fork, we could prove any true arithmetical sentence. 'Making the right choice' in the sense of the theorem is in fact equivalent to choosing (at limit ordinals) a particularly convoluted definition of the axioms of a theory in the sequence, a definition which we know to actually define those axioms only if we already know the sentence which we seek to prove to be true."

We can draw several conclusions from what was stated above. The most trivial one would be that the totality of mathematical truths is not computable, but this was to be expected in the light of Gödel's Theorem. This is not damaging to the Dialetheist position per se, since Priest himself argues that Incompleteness is a necessary product of consistency. The stronger consequence, which does affect the Dialetheist argument is that for no computable ordinal α can T_α prove its own Gödel sentence. So the hope that we could somehow collapse into a recursively enumerable Theory which can achieve the contradiction that Priest is aiming at is futile - Feferman's Completeness Theorem shows that there is such a sequence of theories in the hierarchy, but this sequence depends on the choice we make at each limit stage, and this choice is not computable, as stated by Torkel above. An even stronger consequence can be drawn, in the light of the quote above: that mathematical 'truths' will eventually become a matter of convention, depending on which axioms mathematicians will choose to accept in the future. In other words, past a certain level up to which our choices are guided by intuition, the project of Mathematics will largely be concerned with proving 'If,... then...' statements;

namely "If we choose to accept such and such axiom, then such and such Theorem will follow".

8. Mathematical Practice as a Turing Machine

Our following discussion may be deemed superfluous or vacuous, given that we have already shown that Gödel's Incompleteness Theorem can not be used to argue for the Dialetheist position. It is, however, useful to show that some of the commitments the Dialetheist makes are simply false from the perspective of a modern Theory of Information.

As we have seen above, the Dialetheist argument makes a strong commitment towards the view that our naive proof procedures are recursively enumerable, which in turn, in the light of the Church-Turing Thesis, implies that the practice of Mathematics can be simulated by a Turing Machine. This commitment professor Priest makes without any doubts.

Now, if we are to keep the language of "In Contradiction", this Turing Machine that simulates our naive proof procedures could either be coded "synchronically" or "diachronically". The terminology is easy to understand: the term "synchronic" means there is a "static" underlying set of axioms and rules of inference which is not to be modified at any stage of our proving process; "diachronic" means that at a later stage we will revise our axioms/rules and possibly modify them in the light of empiric considerations, like computational evidence, or even informal intuitions. Either way, it is the Dialetheist position, our naive proof procedures are recursively enumerable, so that Gödel's Theorem can be applied to their theory.

In the words of Priest:

"It is sometimes suggested that proof may not be recursive since we may, from time to time, add to our axioms or rules of proof[...] This cannot per se be used as an argument against the thesis of the recursiveness of proof being defended here, since the canons of proof in question were defined synchronically."

It is useful to develop this point further and try to see whether Priest is commited to a diachronic formalization of naive Mathematics or synchronicity suffices for his purposes. We have already argued above that, assuming Dialetheism would be proven correct, the process of specifying consistent foundations for branches of Mathematics would become diachronic.

It is also useful to note that the only way in wich we could possibly obtain a theory that can prove its own Gödel sentence is by going up the Feferman hierarchy transfinitely, which points to a necessarily diachronic formalization.

Professor Priest then goes on to identify the Gödel sentence as the source of the diachronicity and refer to Feferman's paper, quite correctly. By the end of the discussion, he seems to be prepared to concede the diachronicity of naive proof, however this, he contends, does not affect his argument:

"there is a clear sense in which, whatever the changes are that are made to allow this proof, they are not arbitrary, but are a natural projection of the prior proof procedures", and:

"Even granted that it is the diachronic proof relation which is relevant in this context, there are good reasons for supposing this to be recursive too. As we have noted, the manoevre which is used in transcending the old system is not a random or arbitrary one, but quite a determinate rule-governed one. Thus, on this conception we have not only rules of proof for generating theorems but also rules for generating rules of proof for generating theorems. But theorems in the diachronic sense are still generated by effective rules, and so are recursively enumerable. By Craig's Theorem, the system has a decidable set of axioms, and therefore a recursive proof relation. Indeed, given that this whole process is just as teachable and learnable as the synchronic one, similar considerations will push us to the conclusion that diachronic proof is recursive."

This second quote is particularly contentious. Surely the Dialetheist does not refer to the whole of Mathematics when arguing that its evolution must be entirely rule-governed, that specified, uncontroversial criteria can be applied when choosing which axioms we should next accept, and so on. For this, one can easily refer to the discussion in [14] to see how a verisimilitude like the Axiom of Constructibility can reasonably be doubted.

The quote does, however, contend that a fragment of Mathematics can be (diachronically) formalized such that to provably entail a contradiction. The question then revolves around whether or not this fragment is indeed axiomatizable in a mechanical way or not.

This is in close intimate relationship to the anti-mechanist objection that Priest responded to in the book. The bottom-line question is whether a human Mathematician would have a higher capability in determining mathematical truths than a Turing Machine; if the answer is yes, as Lukas or Penrose or even Gödel himself (a strong believer in the misticism of the human mind) would contend, then arguably the Gödel sentence of the formal system in question would be identified using a "leap of creativity" on the part of the human mathematician, which cannot be formalized, not even in principle.

This, in turn, could be very well explained by a so called "multiple Turing Machines model", which states that the Mathematical practice cannot be simulated by a single Turing Machine, but by several such machines that might overlap in scope, but not entirely. In other words, Mathematics is not to be unified under the umbrella of a single Formal Axiomatic System.

In the words of Turing, as quoted in [8]:

"Let δ be a sequence [e.g. 10111001...] whose n-th figure is 1 or 0 accordingly as n is or is not satisfactory. It is an immediate consequence of [a previous theorem] that δ is not computable. It is (so far as we know at present) possible that any assigned number of figures of δ can be calculated, but not by a uniform process. When sufficiently many figures of δ have been calculated, an essentially new method is necessary in order to obtain more figures." (Turing in 'On Computable Numbers', 1936)

To develop this point further, we must consider that Mathematics as such has a non-algorithmic content, which mathematicians can nevertheless establish through creativity. Consider the following theorem:

Theorem 4 (The Halting Problem). *The Halting Problem is described as following: given a description of a computer program, decide whether the program finishes running or continues to run forever. This is equivalent to the problem of deciding, given a program and an input, whether the program will eventually halt* *when run with that input, or will run forever.*

This problem is undecidable algorithmically.

This result is due to Alan Turing and several other problems can be shown to be undecidable using reductions to the Halting problem.

Among such famous problems for which algorithms cannot exist are the problem of finding integer solutions to diophatine equations, deciding whether or not a set of tiles can tile the plane or deciding equivalences of words in the word problem.

That finding the integer solutions to diophatine equations is undecidable does not prevent the working mathematician from solving many of the problem instances: they could guess the solutions or they could come up with ingenious ways to demonstrate that such solutions cannot exist. This goes to show that the human mind would be by far more creative in the realm of Mathematics than any single Turing Machine could be. Hence the domain of Mathematical Proof itself would be non-recursive as such, since algorithmically undecidable problems could still be decided in principle through the use of creativity.

This presents itself as a problem for Priest, who argues for the view of mathematical practice as a Turing Machine. He is forced into this position because he needs to maintain that our naive proof procedures are recursive, an assumption he needs in order to apply Gödel's First Incompleteness Theorem.

This difficulty for the dialetheist position should be very clear already; the retort, which is present in the book, could be that our Formal Axiomatic System used for formalizing our informal proof procedures would necessarily be coded in a diachronic fashion, with rules for choosing new axioms and rules of inference.

This line of arguing is mistaken at its core: first of all, once we have settled for these meta-rules of the system, there could presumably be a need for transcending this system with meta-meta-rules and so on. Secondly, the distinction between synchronic and diachronic is misleading here, in the sense that there cannot be a quintessentially diachronic system: once the rules have been specified, and the meta-rules, and the meta-meta-rules (we need to stop somewhere) the system will essentially need to be coded in a synchronic fashion!

The next section elaborates on the power of such a Formal Axiomatic System. A good overview of the non-recursive content of Mathematics together with a tangential discussion is given on [3].

9. The Information-Theoretic Argument

We have argued above that the synchronic/diachronic dichotomy is misleading. This is because a "diachronic" Formal Axiomatic System with rules, meta-rules, meta-meta-rules and so on is essentially still equivalent to a Turing Machine, just as a "synchronic" system. As a result, diachronic and synchronic are equivalent notions, and none of them can reasonably account for a leap in creativity that a human mathematician would make when producing a mathematical proof relating to non-algorithmic content. This goes as evidence against the thesis of non-recursiveness of our informal proof procedures, quite against what Priest tries to show.

So we have seen that a Formal Axiomatic System will necessarily have to be coded synchronically. In what follows, we will prove that, once coded, such a FAS will have limited computational power in the sense that there will be problems outside of its computational range. This is the final blow to the thesis of mathematical practice being a Turing Machine, which is a commitment that the dialetheist makes when arguing that informal proof must be recursive.

The following definition is given in [5].

Definition 1 (Halting Probability Ω). *Let us run programs selected randomly on a given fixed computer (Universal Turing Machine). We can effectively do this by flipping a fair coin to decide the next bit of the program. Consider out of all such programs, only the ones that are self-delimiting.*

We then sum for each program that halts, the probability of selecting that program randomly:

$\Omega = \Sigma_{p \in P} 2^{-|p|}$, *where P is the set of programs that halt and $|p|$ is the size of program p in bits.*

The result is the Halting Probability Ω of the given fixed computer, which is machine dependent.

The condition that the programs must be self-delimiting is an important technicality that allows us to sum over all programs of arbitrary size.

The Halting Probability is an important construction, which has coded within it information about whether or not computer programs halt. In the words of Chaitin, with this definition we "get irreducible mathematical facts, which 'are true for no reason', and which simulate in pure math, as much as is possible, independent tosses of a fair coin: the bits of the base two expansion of the halting probability". We can, in effect, use the first N bits of Ω to settle the halting problem for programs up to N bits in size.

Why is this construction relevant or important? Consider the following pseudo-code:

Algorithm 1 Counter-example to Goldbach Conjecture

Require: $\forall i \in N (2i = x + y)$
Ensure: x, y are primes;
  $i = 0$
  while true **do**
    if i is even **then**
      search all pairs of primes (x, y) less than i
      if some of the pairs add to i, CONTINUE, otherwise HALT;
    end if
    $i = i + 1$
  end while

This program will halt if and only if the Goldbach Conjecture is false. Now, if we could somehow compute the Halting Probability of the given machine we are running this program on, it would follow that we can settle the Goldbach Conjecture, since Ω tells us if the program halts or not. In this fashion, we could settle a considerable chunk of mathematical conjectures, provided counter-examples can be coded in this fashion and they refer to countable structures. It goes without saying that Ω is, as a matter of fact, uncomputable, since the Halting Problem is not decidable.

The following result, due to Chaitin, is the key of what we are aiming to show. By the complexity of a Formal Axiomatic System we understand the approximate size of its content of irreducible information, that is, information which cannot be coded in an essentially smaller size.

Theorem 5. *A given Formal Axiomatic System can only determine as many bits of Ω as its complexity. In particular, for any given FAS there is a constant c such that the FAS with complexity $H(FAS)$ can never determine more than $c + H(FAS)$ bits of the value for Ω. The theorems in question are of the form "The 155th bit of Ω is 0" or "The 234th bit of Ω is 1".*

In other words, for any given Formal Axiomatic System (including one that does not involve the principle of explosion, so could tolerate contradictions) there exist infinitely many computational problems that cannot be settled within the system. So the answer to the question to which Priest confessed agnosticism, namely if a dialetheist system could be complete, seems to be 'No'. Hence completeness is unattainable, full stop; this goes to show that incompleteness is an ubiquitos phenomenon, which has nothing to do with whether or not we choose to sacrifice consistency.

But what is even more damaging to the dialetheist argument is the fact that we could perhaps settle arbitrary many bits of Ω in a non-algorithmic fashion by using ingeniousity. We are aware that this, in itself, is a contentious premise, but we choose to accept it as a thesis and will not insist upon arguing for it.

So if indeed it is the case that creativity can add to what we can establish in Mathematics, then it would necessarily follow that our naive proof procedures are, as a matter of fact, non-recursive. If, however, the contrary is true, namely that there cannot exist non-algorithmic creativity, then the FAS would still be limited computationally and the value of the bits of Ω past a certain point would be described by which particular axioms we choose to add to the system, in which case the addition must be arbitrary, hence non-recursive (for if it was recursive, it would have been formal, hence coded in the initial system).

Either way, we can argue strongly that mathematical proofs are non-recursive, essentially due to the very same incompleteness phenomenon discovered by Gödel.

10. Conclusion

If the foundations for our Mathematics are consistent is beyond the limits of what we can ever prove, of course, and the Dialetheist argues for the opposite conclusion. But in order to get there, assuming he could get there in the first place, Priest needs a few missing ingredients: he needs to show beyond any doubt that the axioms/rules of the closed proof system which he deems inconsistent are indeed recursively enumerable. This is deeply problematic, since even for the small fragment of Mathematics known as True Arithmetic, this is known not to be so.

Furthermore, there would be a need for the Dialetheist to draw some boundaries and be explicit in defining the fragment of Mathematics which does entail the inconsistencies. This can not be the whole of Mathematics, since this is not a recursively enumerable theory.

However, the road seems to be blocked from the start: we need to go up the Feferman hierarchy if we aim at a theory which can prove its own Gödel sentence. Whether or not Feferman's Completeness Theorem leaves room for the Dialetheist argument can still be reasonably disputed. Further philosophical interpretations regarding Feferman's Completeness Theorem would be very welcomed and could shed more light on this topic.

All in all, we contend that we have proven that the thesis the dialetheist commits to, that the mathematical practice must be recursive blatantly contradicts basic principles of Algorithmic Information Theory.

Acknowledgements

I must thank Professor Edwin Mares of Victoria University of Wellington Philosophy Department for supervising this project and for many useful comments. Many thanks are also due to Dr David Diamondstone for comments regarding the misuse of Gödel's Incompleteness Theorem in this context. Finally, I am greatly indebted to all of my Math and Philosophy lecturers from which I have learned a great deal in my stay at Victoria University of Wellington.

References

[1] Graham Priest, *In Contradiction. A study of the transconsistent.* Clarendon Press, Oxford, 2nd Edition, 2006.
[2] Graham Priest, Personal communication, 2011.
[3] Roger Penrose, *The Emperor's New Mind. Concerning Computers, Minds, and the Laws of Physics.* Oxford University Press, 1989.
[4] Douglas Hofstadter, *Gödel, Escher, Bach: an eternal golden braid.* New York: Basic Books, 1979.
[5] Gregory Chaitin, *Meta Math! The Quest for Omega.* Vintage Books, New York, 2006.
[6] Solomon Feferman, *Transfinite Recursive Progressions of Axiomatic Theories.* The Journal of Symbolic Logic, Vol. 27, No. 3 (Sep., 1962), pp. 259-316.
[7] Torkel Franzen, *Transfinite Progressions: A Second Look at Completeness.* The Bulletin of Symbolic Logic, Vol. 10, No. 3 (Sep., 2004), pp. 367-389.
[8] Jack Copeland, *The Mathematical Objection: Turing, Gödel, and Penrose on the Mind.* manuscript, July 2008.
[9] S. C. Kleene, *On Notation for Ordinal Numbers.* The Journal of Symbolic Logic, Vol. 3, No. 4 (Dec., 1938), pp. 150-155.
[10] David Diamondstone, Personal communication, 2011.
[11] Alfred Tarski, *The Semantic Conception of Truth and The Foundations of Semantics.* (1944) www.ditext.com/tarski/tarski.html.
[12] Saul Kripke, *Outline of a Theory of Truth.* The Journal of Philosophy Vol. 72, No. 19 (Nov. 6, 1975), pp. 690-716.
[13] Chris Mortensen, *Inconsistent Mathematics.* Kluwer Academic Publishers, Dordrecht, The Netherlands, 2nd Edition, 2010.
[14] Chris Freiling, *Axioms of Symmetry: Throwing Darts at the Real Number Line.* The Journal of Symbolic Logic, Vol. 51, No. 1 (Mar., 1986), pp. 190-200.
[15] Paul Cohen, *Set Theory and the Continuum Hypothesis.* Benjamin, New York, 1966.

School of Mathematics, Statistics and Operations Research, Victoria University, Wellington, New Zealand
E-mail address: valentin.bura@gmail.com

Psychoanalysis is incomplete. An ontological argument

Valentin Bura

June 28, 2023

Abstract

We use Gödel's Incompleteness Theorem to argue no rigorous theory can prove or disprove some statement which psychoanalytical interpretation or free-association may interpret correctly.

1 Preliminaries

This paper is a very brief focus on the following passage. It appears Jung is well aware of the limitations of the psychoanalyst method and so we try to give a formal account of the main idea expressed here.

> Certain dreams, visions or thoughts can suddenly appear; and however carefully one investigates, one cannot find out what causes them. This does not mean that they have no cause; they certainly have. But it is so remote or obscure that one cannot see what it is. In such a case one must wait either until the dream and its meaning are sufficiently understood, or until some external event occurs that will explain the dream.
>
> Carl Gustav Jung, *Man and his Symbols*

Mythos refers roughly to abstract patterns expressing the interconnected web of prevalent beliefs and attitudes in a group or culture, delineated by symbols. A *belief* is the state of mind in which a person or a group think something to be the case, with or without there being empirical evidence to prove that something is the case with factual certainty. An *archetype* is a constantly recurring symbol or motif in the mythos of a group of people. We will refer to archetypes as *archetypal elements*. An *archetypal collection* is a non-empty collection $\mathcal{E} = \{e_1, e_2 \cdots\}$ where each $e_i \in \mathbb{N}$ is a place-holder for an archetypal element. A *belief collection* is a non-empty collection $M = \{m_1, m_2 \cdots\}$ such that each m_i is a place-holder for a given belief. Note the elements that M place-holds are not required to be independent or consistent with each other.

Example 1. *The archetypal elements of the Western Zodiac form an archetypal collection.*

Example 2. *The archetypal elements known as the Traffic Signs form an archetypal collection.*

2 First Order Theory of Integers

Definition 1 (Propositional Calculus). *The propositional calculus PC is defined as follows.*

The symbols of PC are $\neg, \implies, (,)$ and the letters p_i with $i \in \mathbb{N}$;

All statement letters are well formed formulas;

If $\mathcal{A}$ and $\mathcal{B}$ are well-formed so are $\neg \mathcal{A}$ and $\mathcal{A} \implies \mathcal{B}$;

The axioms of PC are

(A1) $(\mathcal{A} \implies (\mathcal{B} \implies \mathcal{A}))$;

(A2) $((\mathcal{A} \implies (\mathcal{B} \implies \mathcal{C})) \implies ((\mathcal{A} \implies \mathcal{B}) \implies (\mathcal{A} \implies \mathcal{B})))$;

(A3) $(((\neg \mathcal{A}) \implies (\neg \mathcal{B})) \implies (((\neg \mathcal{A}) \implies \mathcal{B}) \implies \mathcal{A}))$.

The rule of inference is

(MP) $\mathcal{B}$ is a direct consequence of $\mathcal{A}$ and $\mathcal{A} \implies \mathcal{B}$.

Definition 2 (First Order Language).

The symbols of FOL are the symbols of PC together with the universal quantifier $\forall$ and a collection of function symbols f_i and relation symbols R_i with $i \in \mathbb{N}$, together with an assignment of an arity for every function symbol and every relation symbol.

An *elementary* formula of FOL is a well-formed formula $P(p_1, p_2 \cdots p_t)$ involving propositional letters $p_1, p_2 \cdots p_t$ that does not contain $\neg$.

Definition 3 (Omega). *The natural numbers ω are defined inductively as follows:*

$\emptyset$ is an element of ω,

for any $x \in \omega$, $x + 1 := x \cup \{x\}$ is an element of ω.

The First Order Theory of the Natural Numbers is obtained from FOL and the following axioms.

Definition 4 (Peano Axioms PA).

1. *0 is an element of ω;*

2. *For every $x \in \omega$, $x = x$;*

3. *For every $x, y \in \omega$, $x = y$ implies $y = x$;*

4. *For every $x, y, z \in \omega$, $x = y$ and $y = z$ implies $x = z$;*

5. *For every $x, y \in \omega$, $x = y$ and $x \in \omega$ implies $y \in \omega_E$;*

6. *For every $x \in \omega$, $x + 1 \in \omega$;*

7. *For every $x, y \in \omega$, $x = y$ implies $x + 1 = y + 1$;*

8. *For every $x \in \omega$, $x + 1 \neq 0$;*

9. If κ is a collection such that $0 \in \kappa$, and for every n, if $n \in \kappa$, then $n+1 \in \kappa$, then $\kappa = \omega$.

Definition 5 (Interpretation). *An M-interpretation for a language L is a correspondence between L and a collection M:*

for every n-ary function symbol f of L, a function $f^{\mathcal{M}}: M^n \to M$;

For every n-ary relation symbol R of L, a relation $R^{\mathcal{M}} \subseteq M^n$.

Definition 6 (Structure). *A structure for a language L consists of a non-empty collection M and an M-interpretation of L.*

Structures will also be called *models*.

Lemma 1. *Definition 3 induces a model $\mathcal{N}$, called the* standard model *for the theory of integers.*

Definition 7 (Recursion). *The following functions are recursive:*

The constant: $f(x_1, x_2 \cdots x_t) = n$,

The successor: $f(x) = x + 1$,

The projection: $f_i(x_1, x_2 \cdots x_t) = x_i$,
together with the functions obtained from recursive functions by

Composition: $f \circ (g_1, g_2 \cdots g_t) = f(g_1, g_2 \cdots g_t)$,

Recursion: $g \otimes h := f(y, x_1, x_2 \cdots x_t)$ where $f(0, x_1, x_2 \cdots x_t) = g(x_1, x_2 \cdots x_t)$ and $f(y+1, x_1, x_2 \cdots x_t) = h(y, f(y, x_1, x_2 \cdots x_t), x_1, x_2 \cdots x_t)$.

A recursive function will be called *computable*.

3 Archetypal Theory

We define the archetypal ordinal omega by iteration of the union operator, starting from a base archetypal collection. A writing system based entirely on logograms/logographs, a pure logographic system, such as Toki Pona, is an example of an archetypal omega. Such a system is relative to the collection of its morphemes.

Definition 8 (Archetypal Omega). *Let $\mathcal{E}$ be an archetypal collection. The archetypal omega of $\mathcal{E}$ is a collection $\omega_{\mathcal{E}}$ defined inductively as follows:*

$\mathcal{E}$ is an element of $\omega_{\mathcal{E}}$,

if $x \in \omega_{\mathcal{E}}$, $x + 1 := x \cup \{x\}$ is an element of $\omega_{\mathcal{E}}$.

An *archetypal theory* $\Psi(\omega_{\mathcal{E}})$ is a collection of sentences of FOL together with the following axioms.

Definition 9 (The Archetypal Axioms). *Let $\mathcal{E}$ be an archetypal collection.*

1. $\mathcal{E}$ is an element of $\omega_{\mathcal{E}}$;
2. For every $x \in \omega_{\mathcal{E}}$, $x = x$;
3. For every $x, y \in \omega_{\mathcal{E}}$, $x = y$ implies $y = x$;

4. For every $x, y, z \in \omega_{\mathcal{E}}$, $x = y$ and $y = z$ implies $x = z$;

5. For every $x, y \in \omega_{\mathcal{E}}$, $x = y$ and $x \in \omega_{\mathcal{E}}$ implies $y \in \omega_{\mathcal{E}}$;

6. For every $x \in \omega_{\mathcal{E}}$, $x + 1 \in \omega_{\mathcal{E}}$;

7. For every $x, y \in \omega_{\mathcal{E}}$, $x = y$ implies $x + 1 = y + 1$;

8. For every $x \in \omega_{\mathcal{E}}$, $x + 1 \neq \mathcal{E}$;

9. If κ is an archetypal collection such that $\mathcal{E} \in \kappa$, and for every archetypal collection α, if $\alpha \in \kappa$, then $\alpha + 1 \in \kappa$, then $\kappa = \omega_{\mathcal{E}}$.

An *archetypal language* $\mathcal{L}_{\mathcal{E}}$ is the language of an archetypal theory.

Definition 10 (Archetypal Interpretation). *An* archetypal interpretation *for an archetypal language $\mathcal{L}_{\mathcal{E}}$ is a correspondence between $\mathcal{L}_{\mathcal{E}}$ and a belief collection M:*

for every n-ary function symbol f of $\mathcal{L}_{\mathcal{E}}$, a function $f^{\mathcal{M}}: M^n \to M$;

for every n-ary relation symbol R of $\mathcal{L}_{\mathcal{E}}$, a relation $R^{\mathcal{M}} \subseteq M^n$.

Definition 11 (Free Association). *A free association for an archetypal language $\mathcal{L}_{\mathcal{E}}$ is an M-interpretation for $\mathcal{L}_{\mathcal{E}}$ where $M \subseteq M_1 \times M_2 \times \cdots M_t$ and each M_i is a belief collection.*

Definition 12 (Archetypal Structure). *A archetypal structure $\mathcal{A}$ for an archetypal language $\mathcal{L}_{\mathcal{E}}$ consists of a non-empty belief collection M and:*

an M-interpretation of $\mathcal{L}_{\mathcal{E}}$, or

a free-association of $\mathcal{L}_{\mathcal{E}}$.

An archetypal structure will also be called *archetypal model*.

Lemma 2. *Definition 8 induces an archetypal model for the archetypal theory.*

Lemma 3. *Any archetypal omega is recursive.*

Proof. Follows from Definition 7 and Definition 8. □

4 Cryptography

Definition 13 (Encryption). *An* encryption *is a bijective map $\epsilon: W \to M$ where W is a collection of* words *and M is a belief collection.*

An encryption is information-theoretically secure if its security derives purely from information theory. That is, it cannot be broken even when the adversary has unlimited computing power.

Definition 14 (Secure Encryption). *An encryption ϵ is secure if ϵ^{-1} is not computable.*

5 Universal Grammar

Definition 15. *An* archetypal production rule *is a pair* $(s, s') := \{\{s\}, \{s, s'\}\}$, *where* $s, s' \in \omega_{\mathcal{E}}$.

Remark 4. $\omega_{\mathcal{E}}$ *contains all its production rules.*

Definition 16 (Universal Grammar). *Let $\mathcal{E}$ be an archetypal collection. A* grammar *of $\mathcal{E}$ is a collection of production rules $\Gamma \in \omega_{\mathcal{E}}$. The* universal grammar $\Gamma_{\mathcal{E}}$ *is the totality of production rules of $\mathcal{E}$.*

Remark 5. *Let $\mathcal{E}$ be an archetypal collection. Then, $\Gamma_{\mathcal{E}} \cong \omega_{\mathcal{E}}$.*

Proof. Let $\pi \colon \Gamma_{\mathcal{E}} \to \omega_{\mathcal{E}}$, $\pi(s, s') = 1/2(s + s')(s + s' + 1) + s$, and define isomorphism ι by: $\mathcal{E} \mapsto \pi_t^{-1}(\mathcal{E})$ and $e \mapsto \pi_t^{-1}(e)$. $\square$

6 Logical Atomism

In Tractatus the following considerations are made:

    i Every proposition has a unique final analysis which reveals it to be a truth-function of elementary propositions (3.25, 4.221, 4.51, 5);

    ii The elementary propositions assert the existence of atomic states of affairs (3.25, 4.21);

    iii Elementary propositions are mutually independent - each one can be true or false independently of the truth or falsity of the others (4.211, 5.134);

    iv Elementary propositions are immediate combinations of semantically simple symbols or "names" (4.221);

    v Names refer to items wholly devoid of complexity, denoted as "objects" (2.02 & 3.22);

    vi Atomic states of affairs are combinations of these simple objects (2.01).

These motivate the following terminology. Let $\mathcal{E}$ be an archetypal collection. A *name* ν is an element of ω_E. An *elementary proposition* $\bar{p}(\nu_1, \nu_2 \cdots \nu_t)$ is an elementary formula $P(p_1, p_2 \cdots p_t)$ of FOL where p_i is a place-holder for the name ν_i. A collection of propositions is a collection $\bar{\xi}$ of elementary propositions. The negation operator $N(\bar{\xi})$ is defined as $N(\bar{\xi}) := \{\neg \bar{p} \mid \bar{p} \in \bar{\xi}\}$. The collection of composite propositions $[\bar{p}, \bar{\xi}, N(\bar{\xi})]$ is defined as follows:

$\bar{p} \in [\bar{p}, \bar{\xi}, N(\bar{\xi})]$ where $\bar{p}$ is an elementary proposition,

$N(\bar{p}) \in [\bar{p}, \bar{\xi}, N(\bar{\xi})]$,

$P(\bar{p}_1, \bar{p}_2 \cdots \bar{p}_t) \in [\bar{p}, \bar{\xi}, N(\bar{\xi})]$.

We note that (v) above defines "objects" as "devoid of complexity", hence our account of the archetypal omega *is not* atomistic.

7 Theoretical Implications

Remark 6 (Gentzen's Consistency Theorem). *The system $FOL + PA$ is consistent.*

This is the celebrated result of Gödel.

Remark 7 (Gödel's Incompleteness Theorem). $\exists G \in L(FOL + PA)$, $FOL + PA \nvdash G \wedge FOL + PA \nvdash \neg G$.

Rosser's formulation is the following.

Corollary 8 (Rosser). *Let T be a consistent formalized system and let $PA \subseteq T$. Then: $\exists R_T \in L(T)$ such that $T \nvdash R_T \wedge T \nvdash \neg R_T$.*

It is provable that the standard model of integers is a model for the archetypal theory.

Lemma 9. $\mathcal{N} \models \Psi$

Proof. Define isomorphism ι by: $0 \mapsto \mathcal{E}$ and let $x + 1 \mapsto \alpha + 1$, for $x \in \omega$ and $\alpha \in \omega_\mathcal{E}$ such that $\iota(x) = \alpha$. Then, by using nested induction on n, m we can establish that:

$$\iota(n) \in \omega_\mathcal{E} \text{ iff } n \in \omega;$$
$$\iota(n + m) = \iota(n) + \iota(m);$$
$$\iota(n) = \iota(m) \text{ iff } n = m.$$

□

The implication is that the model of integers is isomorphic to any archetypal structure.

Corollary 10. $\mathcal{N} \cong \mathcal{A}$

Proof.

    i. if the archetypal structure is an archetypal interpretation, the result follows from Lemma 9;

    ii. if the archetypal structure is a free association, let $\pi \colon \mathbb{N}^2 \to \mathbb{N}$, $\pi(n, m) = 1/2(n + m)(n + m + 1) + m$, let $\pi_t \colon \mathbb{N}^t \to \mathbb{N}$, $\pi_t(n_1, n_2 \cdots n_t) = \pi(\pi_(t - 1)(n_1, n_2 \cdots n_{t-1}), n_t)$, and define isomorphism ι by: $0 \mapsto \pi_t^{-1}(0)$ and $n \mapsto \pi_t^{-1}(n)$.

□

Remark 6 and Corollary 10 imply the fragment of psychoanalytical theory defined hereby is consistent.

Corollary 10 and Remark 7 imply psychoanalytic theory is incomplete in the following sense: there exist statements of its language which may be interpreted or freely-associated correctly that an archetypal theory cannot prove or disprove.

Theorem 11. $\exists \psi \in \mathcal{L}_\mathcal{E}$, *such that* $\Psi \nvdash \psi \wedge \Psi \nvdash \neg \psi$.

The following result states there exists no complete archetypal theory of a Universal Grammar.

Corollary 12 (Private Language Argument). *Let $\mathcal{E}$ be an archetypal collection. $\exists \psi \in \mathcal{L}_\mathcal{E}$, such that $\Psi(\Gamma_\mathcal{E}) \nvdash \psi \wedge \Psi(\Gamma_\mathcal{E}) \nvdash \neg \psi$.*

Proof. Follows from Theorem 11 and Remark 5.

□

8 Further Implications

For a given formula A of $FOL + PA$, the encoding $g(A)\colon L \to \omega$ and the predicate $T(A, g(A))$ which holds if $g(A)$ is an encoding for A are defined in [2]. The formula $G(A)$ is then defined as $A(g(\neg A))$ and the predicate $CON(S)$ as $\neg \exists A, S \vdash A \land S \vdash \neg A$.

Remark 13 (Tarski). *There exists no L formula $True$ of $FOL+PA$ such that $True(A) \iff A$.*

Proof. Let $g(A)$ be the encoding of A. Let $True(A) := \exists g(A) \in \omega, T(g(A)) \iff A$. □

This implies the truth-predicate is not definable in the psychoanalytical theory.

Corollary 14. *Let $\mathcal{E}$ be an archetypal collection. Then there exists no formula $True$ of $\Psi(\omega_\mathcal{E})$ such that $True(A) \iff A$*

Proof. Follows from Remark 13 and Corollary 10. □

Corollary 15. *There exists an archetypal collection $\mathcal{E}$ and a collection $S_\mathcal{E} \in \mathcal{L}_\mathcal{E}$ such that $S_\mathcal{E}$ is secure.*

Proof. Follows from Theorem 11 and Definition 7 by the following construction of $S_\mathcal{E}$:

for any formula A provable in Ψ, let $G(A) \in S_\mathcal{E}$,

for any formula A of $S_\mathcal{E}$, let $G(A) \in S_\mathcal{E}$.
□

Remark 16 (Gödel's Second Incompleteness Theorem). $\mathcal{N} \nvdash CON(\mathcal{N})$.

As an implication, we note that no archetypal theory can prove its own consistency.

Corollary 17. *Let $\mathcal{E}$ be an archetypal collection. Then, $\Psi(\omega_\mathcal{E}) \nvdash CON(\Psi(\omega_\mathcal{E}))$.*

Proof. Follows from Corollary 10. □

Remark 18 (Gödel's Completeness Theorem). *Every consistent, countable first-order theory has a finite or countable model.*

For a given collection of sentences Σ, let $Mod(\Sigma) := \{\mathcal{M} \mid \mathcal{M} \models \Sigma\}$.

Remark 19 (Compactness Theorem). *Let L be a language.*

1. *For all collections Σ and Π of L-sentences, then $\Sigma \vdash \Pi$ if and only if $Mod(\Sigma) \subseteq Mod(\Pi)$.*

2. *If T is an L-theory, then $Mod(T)$ is non-empty.*

In other words, a collection of sentences is consistent if it has a model and a collection of sentences Σ implies a sentence ψ if every model of Σ is also a model of ψ. Hence the compactness theorem can be summarised as follows:

A collection of sentences implies a sentence if and only if it finitely implies the sentence.

A collection of sentences is consistent if and only if it is finitely consistent.

Corollary 20. *The following are equivalent:*

Σ *is a complete L-theory;*

There exists an L-structure $\mathcal{M}$ such that $\mathcal{M} \in Mod(\Sigma)$.

Corollary 21. *Let Ψ be an archetypal theory. For any finite $\Sigma \in \Psi$ there exists an archetypal structure $\mathcal{S}$ such that $\mathcal{S} \models \Sigma$.*

Proof. By Remark [6], Lemma [2] and Corollary [20], the archetypal theory has an archetypal structure. □

Corollary 22. *Let Ψ be an archetypal theory. Then there exists an archetypal structure $\mathcal{A}$ such that $\mathcal{A} \models \Psi$.*

Proof. Follows from Corollary [21] and Corollary [20]. □

References

[1] Carl Gustav Jung, *Man and his Symbols*, 1964.

[2] Kurt Gödel, *On formally undecidable propositions of Principia Mathematica and related systems*, 1931.

[3] Gerhard Gentzen, *Untersuchungen über das logische Schließen*, 1935.

[4] Stephen Kleene et al, *Introduction to metamathematics*, 1952.

[5] Ludwig Wittgenstein, *Tractatus Logico-Philosophicus*, 1921.

www.ingramcontent.com/pod-product-compliance
Lightning Source LLC
Chambersburg PA
CBHW062209220526
45470CB00009B/2988